The Long Roll

Wartime Experiences of the
Civil War Drummer Boy

Dedicated to the courageous youth who
went off to a man's war in service to their country.

The Long Roll

Wartime Experiences of the Civil War Drummer Boy

By Michael Aubrecht

Foreword by Daniel Glass

Heritage Books
2025

HERITAGE BOOKS
AN IMPRINT OF HERITAGE BOOKS, INC.

Books, CDs, and more—Worldwide

For our listing of thousands of titles see our website
at
www.HeritageBooks.com

Published 2025 by
HERITAGE BOOKS, INC.
Publishing Division
5810 Ruatan Street
Berwyn Heights, MD 20740

International Standard Book Number
Paperbound: 978-0-7884-4648-1

Contents

Foreword

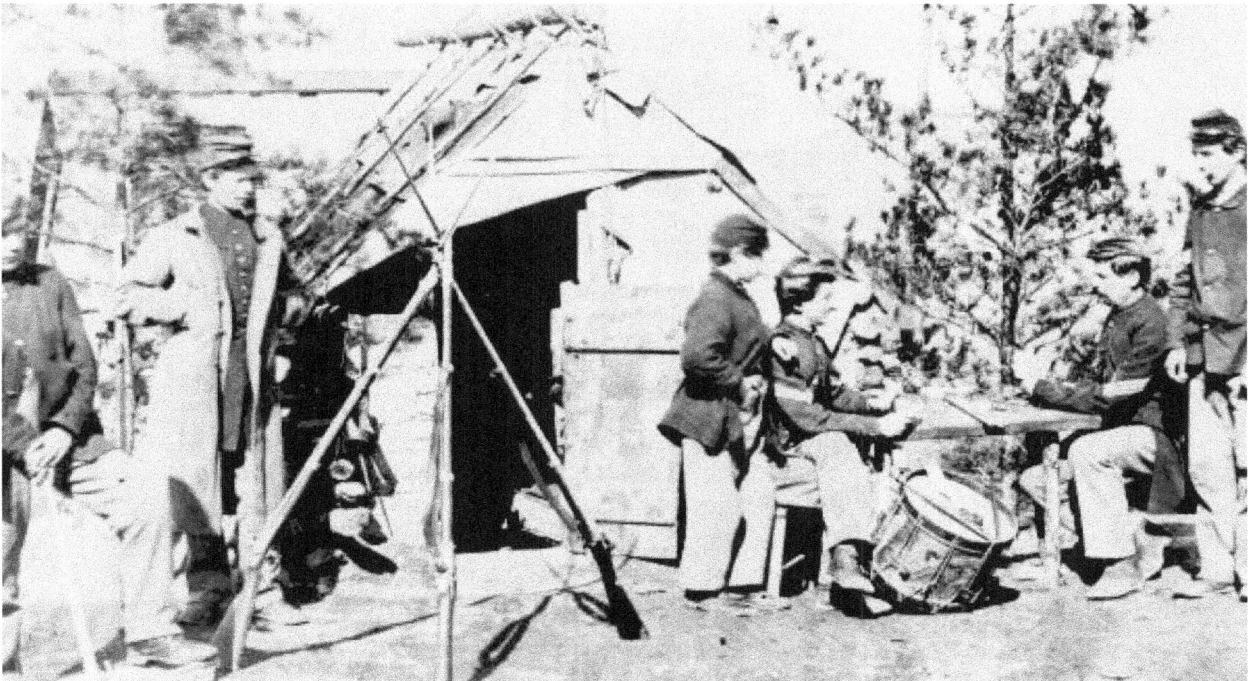

I was first introduced to Michael Aubrecht about five years ago by my pal Rich Redmond; the two were co-authoring a book for young drummers entitled *FUNdamentals of Drumming for Kids*. While I was immediately impressed by Michael's facility behind a drum set, I soon discovered that he was also an acute student of military history, and an avid researcher who had authored several books on the subject. As a researcher and writer with a similar passion for history, I found in Michael a kindred spirit, and we immediately bonded.

Michael's latest project, *The Long Roll*, allows him to fuse both of his long-term interests - drumming and military history - into a fascinating and detailed account of the Civil War Drummer Boy. Like most people, I've seen depictions of young military drummers in paintings and literature throughout my entire life. Every holiday season, I memorialize "the little drummer boy" in song. And yet, as I read through *The Long Roll*, I realized just how little I actually knew about this subject – and how much of a story there was to tell.

Aubrecht lives in Fredericksburg, Virginia - surrounded by the Civil War battlefields of Fredericksburg, Spotsylvania, Chancellorsville and The Wilderness. This proximity has given him a highly personal perspective on his subject matter, as well as firsthand access to unique research opportunities. Through interviews, press clippings, letters and a trove of fascinating photos, Aubrecht humanizes his subject and paints a moving portrait of the incredible courage, dedication and spirit that drummer boys (and on the rare occasion, girls) brought to their job.

From the founding of the Republic through the end of the 19th Century, drummers played a vital role within the apparatus of the United States Military. Beyond providing cadences that kept soldiers on the move during long marches, military drummers also functioned as communicators who provided specific "calls," telling the men when to assemble, when to eat, when to break camp, and when to attack on the battlefield.

The job of drummer was typically held by young men, brimming with a desire to serve, but not yet of the requisite age to do so. Previous musical experience was not always a requirement – many drummer boys (perhaps fueled by dreams of travel, adventure and glory), developed their skills "on the job."

While history has often cast drummer boys in a romantic light – as cherubic "mascots" whose participation in the action was only corollary - Aubrecht offers a far more unvarnished illustration. Many drummer boys were runaways, often seeking respite from a troubled home life. Many falsified their ages when mustering in, meaning that sometimes "… these boys who marched off to participate in a man's conflict" were as young as 14.

Drummer boys did not carry weapons, yet they were expected to be on the front lines supplying calls at the outset of a skirmish. According to Aubrecht, "The precarious risk at which they put themselves was often equal to that of their elders. When not acting in the role of musicians, they often served as stretcher bearers, witnessing firsthand the horrors of war and the carnage it inflicted on those who fell." In this capacity, drummer boys themselves were at great risk of injury or death, and many who survived the killing fields of America's bloodiest conflict no doubt suffered the same kinds of trauma as "shell shocked" soldiers.

With *The Long Roll*, Aubrecht offers a new perspective on an iconic subject. I'd like to be one of the first to congratulate him on this excellent and much needed addition to the larger discussion of American Military Drumming.

Daniel Glass
New York City
May, 2019

Visit Daniel online at www.danielglass.com

From the Author

First, thank you for taking the time to read this book. I've wanted to write a study on the history and experiences of the Civil War Drummer Boy for quite some time. Over the last few years I've written dozens of posts on my blog, collected nearly 100 images, and acquired many first-hand accounts. If you Google "Civil War Drummer Boy" one of my pieces is the third one listed.

Clearly I felt that I had enough material to create something good if I could organize it in a way that's worthwhile to the reader. I anticipated a 30 or so page composition. I reached 50 pages. There were several sections that I envisioned for the book. A dramatic introduction, overview of the drummer boy's origins, stories of noteworthy individuals and quoted first-hand accounts.

There were some challenges. I didn't want to write an academic book although I wanted it to be educational. I wanted it to be visually pleasing but not too "artsy." And as a drummer I needed to refrain from getting technical. Once I accepted those requirements I was able to move forward.

I can promise you there will be a wide range of topics covered in this book. I tried to present the drummer from a variety of perspectives. There will be some debunking of some myths that even I fell for, as well as some stories of individuals that have been long forgotten. The end goal is to give credit to a far too neglected individual in Civil War memory, boys who left home to participate in a man's war. When I wrote my books on Confederate Campsites in Spotsylvania County and the Historic Churches of Fredericksburg I came away with a whole new understanding of the plight of the everyday soldier. I hoped to have the same experience while writing this book. Now upon its completion I can say that I've had a similar understanding.

So here is my offering on the story of the Civil War Drummer Boy. That said, not everyone appreciates my efforts. Here is an interesting period quote I found that presents a take on the stories about drummer boys that isn't too favorable:

Marvelous stories are still related of "drummer boys." I see they are very ridiculous. I have never yet saw a dead drummer boy—not even a wounded one. The little rascals take care and remain far enough in the rear, and are really of no use during an active campaign except to carry water to the hospital. Few of them do this, being employed chiefly in "going down on knapsacks" that are thrown away by wounded soldiers. Drum corps have more mischief in them than in all regiments besides. – William H. Moody, 139th Pennsylvania Volunteers.

Thank you for the inspiration Mr. Moody.

Visit me online at www.michaelaubrecht.wordpress.com

Introduction

As morning broke shadows awoke from their twilight slumber and began to stretch their limbs in acknowledgment of the recurring day. Below in the valley, an army was also just beginning to stir. Many soldiers however, did not share nature's sentiments in welcoming back another sunrise. Exhausted, homesick and terribly traumatized by the horrors they had witnessed on the battlefield, the promise of another day was nothing more than prolonged suffering. After all, weeks had turned into months, months had turned into years, and no end appeared in sight. Many felt as if they had been on campaign forever. Most were only able to find a sense of peace and comfort while sleeping. That is, when they could sleep.

Looking more dead than alive, they were now faded memories of the vibrant men they had once been. Long gone was the patriotism and thrill of recruitment parades and brand new uniforms. No longer were they believers in the promise of adventure or the romance of war. Emerging from their weathered tents, some struck fires as the smell of stale coffee began to permeate the air. The gentle sounds of the surrounding countryside gave way to the neighing of irritated horses. As they began their daily rituals, muskets were inspected, swords were sheathed and once pristine jackets were pulled on over dirty white shirts and tattered suspenders.

The stillness of the morning was broken by the sound of a long roll acting as reveille calling the men to attention. Ironically it was the responsibility of boys to command these men to muster. Boys who had marched off to participate in a man's conflict. The army relied on the services of these boys as musicians and as communicators. Just like their counterparts, they were regulars. Military divisions had multiple drummers spread throughout their ranks. They suffered the same hardships as the men.

Whether playing a monotonous cadence to keep men moving while on long marches, long rolls to call men to assemble in the mornings, or booming signals to communicate for their officers on the battlefield the skill at which the drummers played was far too often overlooked. The courage at which this had to be done was also neglected as the youthful age of these boys paled in comparison to the men they served. The precarious risk at which they put themselves in was often equal to that of their elders and when they were not acting in the role of musician they served as stretcher bearers witnessing firsthand the horrors of war and the carnage it inflicted on those who fell.

As rows of anxious soldiers took the field drummers played the Call to Battle to keep them assembling and in step. Lining up the rank and file the daring infantry waited for the signal to move forward. Standing by the officers on the field the drummer boys managed to maintain their composure despite their obvious fear. They provided critical communications to gesture movement. The scent of smoke filled the air and permeated their already dusty clothes. Once fully engaged they often moved to the rear, exchanging their drums for makeshift stretchers. From then on they struggled to maintain their composure as they carried their bloody comrades off the field. Perhaps that was their greatest challenge of all. Whether drumming on the march or bearing the wounded these courageous boys quickly grew up in a man's war. They had to.

Origins of the Drummer Boy

Throughout the history of warfare musicians have always played an important role on the battlefield. Military music has served many purposes including marching-cadences, bugle-calls and funeral dirges. Fifes, bagpipes and trumpets are just some of the instruments that were used to instruct friend and intimidate foe.

Perhaps the most notable of these instruments was the drum. From as far back as the ancient days of Babylon, the beating of animal skins rallied the troops on the field, sent signals between the masses, and scared the enemy half to death. During the Revolutionary War, drummers in both the Continental and English ranks marched bravely into the fight with nothing but their sticks to protect them.

According to Glenn Williams, a senior historian at the U.S. Army Center of Military History, "Drummer boys signaled everything from chow time to formation to pay to time to charge the enemy. Drummers and other musicians like fifers and buglers were the radio operators of their day. For much of military history, this was the most effective way for commanders to relay their orders to hundreds or even thousands of troops on the battlefield. They also were used to intimidate the enemy. Before the actual battle began, it wasn't unusual for fifes and drums to march out in front of the regiment and play a tune before they fell in with their companies and maybe intimidate the enemy a little bit."

Remarkably, some of these musicians were in fact, very young boys, not quite yet into their teen years. That group however, was a minority. Despite popular culture's portrayal of the little "Drummer Boy," boys were actually an acceptation to the rule in early American warfare. According to *The Music of the Army... An Abbreviated Study of the Ages of Musicians in the Continental Army* by John U. Rees (Originally published in The Brigade Dispatch Vol. XXIV, No. 4, Autumn 1993, 2-8.):

Boy musicians, while they did exist, were the exception rather than the rule. Though it seems the idea of a multitude of early teenage or pre-teenage musicians in the Continental Army is a false one, the legend has some basis in fact. There were young musicians who served with the army. Fifer John Piatt of the 1st New Jersey Regiment was ten years old at the time of his first service in 1776, while Lamb's Artillery Regiment Drummer Benjamin Peck was ten years old at the time of his 1780 enlistment. There were also a number of musicians who were twelve, thirteen, or fourteen years old when they first served as musicians with the army.

Sixteen years, although young by today's standards, was considered the mature age of a young man in the days of the American Revolution. It was also the average age of many fifers and drummers who volunteered to march in the ranks of General George Washington's Continental Army. For example the 11th Pennsylvania Regiment boasted the following musician's roll:

John Brown, fife – 14 years-old (enlisted in 1777), Thomas Cunningham, drum – 18 years-old (enlisted in 1777), Benjamin Jeffries, drum – 15 years-old (enlisted in 1777), Thomas Harrington, drum – 14 years-old (enlisted in 1777), Samuel Nightlinger, drum – 16 years-old (enlisted in 1777), James Raddock, fife – 16 years-old (enlisted in 1777), George Shively, fife – 19 years-old (enlisted in 1777), David Williams, drum – 17 years-old (enlisted in 1777).

Despite their non-combatant roles in battle, many of these drummer's war stories are even more compelling than those of the fighting men around them. For instance, Charles Hulet, a drummer in the 1st New Jersey…The following deposition was given by Hulett's son-in-law in 1845:

"… said Hulett… enlisted in Captain Nichols company [possibly Noah Nichols, captain in Stevens' Artillery Battalion as of November 9, 1776. In 1778 he was a captain in the 2nd Continental Artillery. See entry for Joseph Lummis] which was a part of the first Regiment of New Jersey in the service of the United States which Regiment was commanded by Col. Ogden. He enlisted as aforesaid on the 7 May 1778… He was engaged in the battle of Monmouth and was wounded in the leg and then or soon after taken a prisoner and by the enemy and carried in captivity to the West Indies, To relieve himself from the horrors of his imprisonment he joined the British Army as a musician and was sent to the United States. That soon after his return… he deserted from the British ranks and again joined the army of the United States and the south under General Greene. He was present at the siege of York and after the surrender of Cornwallis he was one of the corps that escorted the prisoners which was sent to Winchester… and he remained in service to the end of the war. This declarant always understood that said Hulett at the close of the war held the rank of Drum-Major."

As primarily noncombatants, it is rare to have a detailed look at the service of any military musician. John George is an exception to that rule as he served the Continental Army's supreme commander as his personal percussionist. His descendants have also done an exceptional job keeping his legacy alive through public commemorations. Arville L. Funk's study titled *From a Sketchbook of Indiana History*, includes a profile of the first "famous" American drummer. It reads:

SGT. JOHN GEORGE
★ ★ BORN 1758 – DIED 1842 ★ ★
DRUMMER BOY IN THE
REVOLUTIONARY WAR FOR
GEORGE WASHINGTON
IN OGDEN'S COMPANY
N.J CONTINENTAL LINE

In a little known grave in south-western Marion County, Indiana, lie the remains of an old soldier traditionally acclaimed as "George Washington's drummer boy." This is the grave of Sergeant John George, a Revolutionary War veteran of the First Battalion of the New Jersey Continental Line. Through extensive and alert research by Chester Swift of Indianapolis into Revolutionary war records, muster rolls, field reports, pension records, etc., there is evidence that Sergeant George might have been the personal drummer boy of Washington's Headquarters Guard during a large portion of the Revolutionary War...On September 8th of that year, Private George, who was listed on the company's rolls as a drummer, fought in his first battle, a short engagement at Clay Creek, which was a prelude to the important Battle of Brandywine. Later, Ogden's battalion was to participate in the battles of Germantown and Monmouth, serving as a part of the famous Maxwell Brigade. The Maxwell Brigade served during the entire war under the personal command of General Washington and was considered to be one of the elite units of the American army. According to John George's service records, he served his first three-year enlistment as a private and a drummer with the brigade at a salary of $7.30 a month. When his three-year enlistment expired, George reenlisted as a sergeant in Captain Aaron Ogden's company of the First Battalion (Maxwell's Brigade) for the duration of the war.

Drummer boys during the American Civil War were younger than their predecessors, but more advanced in their playing. Each drummer was required to play variations of the 26 rudiments. The rudiment that meant attack was a long roll. The rudiment for assembly was (777 flam flam 777 flam flam 777 flam flam 77 flam flam 7) and the rudiments for the drummers call was (7 flam flam 7 flam flam 7 flam flam 2x fast, 1x slow 7 flam 7 flam). The rudiment for simple cadence was (open beating) (55 flam flam repeat). Additional requirements included the double stroke roll, paradiddles, flamadiddles, flam accents, flamacues, ruffs, single and double drags, ratamacues, and sextuplets.

The repertoire of the Civil War drummer boy included "Three Camps" which was reveille, "Tattoo" which meant bedtime and "Commence Firing," "Quick Step," "Advance" and "Retreat" which were all battlefield commands.

Many drummers learned how to play by attending the Schools of Practice at Governor's Island, New York Harbor, and Newport Barracks, Kentucky, although the vast majority learned in the field. Some were aided by texts; the most popular by far was Bruce and Emmett's *The Drummers' and Fifers' Guide*.

According to historian Ron Engleman:

The word rudiments first appeared in a drum book in 1812. On page 3 of A New Useful and Complete System of Drum Beating, Charles Stewart Ashworth wrote, Rudiments for Drum Beating in General. Under this heading he inscribed and named 26 patterns required of drummers by contemporary British and American armies and militias. The word Rudiment was not used again in US drum manuals until 1862. George B. Bruce began page 4 of Bruce and Emmett's Drummers and Fifers Guide with the words Rudimental Principles.

Beginning with the long roll, Bruce listed 35 patterns concluding with a paragraph titled Recapitulation of the Preceding Rolls and Beats. On page 7 of his 1869 Drum and Fife Instructor, Gardiner A. Strube wrote, The Rudimental Principles of Drum – Beating, and followed with 25 examples, each named Lesson.

Military drums were usually 18" or more prior to the Civil War when they were shortened to 12"-14" deep and 16" in diameter in order to accommodate younger (and shorter) drummers. Ropes were joined all around the drum and were manually tightened to create tension that stiffened the drum head, making it playable. The drums were hung low from leather straps necessitating the use of traditional grip. Regulation drumsticks were usually made from Rosewood and were 16"-17" in length. Ornamental paintings were very common for Civil War drums which often depicted pictures like Union eagles and Confederate shields.

The U.S. Army ordered approximately 32,000 drums over the course of the conflict. The younger the drummer, the more difficulty one would have lugging around these cumbersome instruments. However, that aspect didn't deter boys from taking up the instrument. According to a brief history on Civil War drummers:

Although there were usually official age limits, these were often ignored; the youngest boys were sometimes treated as mascots by the adult soldiers. The life of a drummer boy appeared rather glamorous and as a result, boys would sometimes run away from home to enlist. Other boys may have been the sons or orphans of soldiers serving in the same unit. The image of a small child in the midst of battle was seen as deeply poignant by 19th-century artists, and idealized boy drummers were frequently depicted in paintings, sculpture and poetry. Often drummer boys were the subject of early photographic portraits.

The youngest soldier killed during the entire American Civil War (1861–1865) was a 13 year-old drummer boy named Charles King. He had enlisted as a drummer boy in the 49th Pennsylvania Volunteer Infantry with the reluctant permission of his father. On September 17, 1862 at the Battle of Antietam he was mortally wounded near or in the area of the East Woods, carried from the battlefield to a nearby field hospital where he died three days later.

Twelve year-old Union drummer boy William Black was the youngest recorded person wounded in battle during the American Civil War. One of the most famous drummers was John Clem, who had unofficially joined a Union Army regiment at the age of nine as a drummer and mascot. Young "Johnny" became famous as the "The Drummer Boy of Chickamauga" where he is said to have played a long roll and shot a Confederate officer who had demanded his surrender.

On the southern side an 11 year-old drummer in the Confederate Kentucky Orphan Brigade, known only as "Little Oirish," was credited with rallying troops at the Battle of Shiloh by taking up the regimental colors at a critical moment and signaling the reassembly of the line of battle.

Another noted drummer boy was Louis Edward Rafield of the 21st Alabama Infantry, Co. K, known as the "Mobile Cadets." He had enlisted at age 11 and while 12 at the Battle of Shiloh he somehow lost his drum; he then obtained an enemy drum and kept on going, thus earning the title of "The Drummer Boy of Shiloh." (Johnny Clem was also known by "Johnny Shiloh" for a period.)

As the years passed the drum was eventually replaced on the battlefield in favor of the bugle although it often returned during Veteran Reunions that took place decades after the war. Many drummers had gone on to become drum majors or drum instructors themselves.

One notable post-war percussionist was Alexander Howard Johnson, the drummer boy of the 54th Massachusetts Infantry made famous by the film Glory. Johnson was said to be a talented drummer who served with distinction and later became a sought after instructor. According to Johnson, he continued to play the drums all his life.

Four years after the South's surrender Johnson organized "Johnson's Drum Corps." where he led the band as drum major, and styled himself as "The Major." According to a local writer who interviewed Johnson years later.

He is probably one of the best drummers in Massachusetts, and boasts that there is hardly a drummer who marches the streets of Worcester who has not received instruction from him.

No one can dispute the service of the drummer boy who left the safety of their homes and firesides to serve their respective cause in a man's war. It is unfortunate that their contribution is so often overlooked. Their legacy lives on. Today America's armed forces boast some of the most talented musicians in the country with many of them still playing traditional instruments and cadences. Military drummers are still highly respected and carry on the tradition of their instrument's storied history that should not be forgotten.

Courage and Distinction

Photographs of drummer boys captured their youthful appearance. Some of these boys looked to be so young they couldn't have been near their teens. Many of them ran away from home with romantic dreams stuck in their heads of the glorious life of a drummer. No doubt as they witnessed the horrors of war their starry-eyed inclinations were stripped away. One can only imagine the hardships they endured.

There were age-limits to who could enlist but they were often ignored. The youngest enlistees served the roll as "mascot" until they were old enough to serve as a drummer. The stark reality is that most of these boys suffered and endured the same hardships along with the adults. They too marched for miles, fought boredom in camp and performed under fire.

Far too often the efforts of those who don't fit under the title of a "traditional" soldier to include drummers, cooks, teamsters and even horses go unnoticed. The truth of the matter is that none of the soldiers who dominate our memories would have been able to fight without the dedication and efforts of the aforementioned. Drummer boys were responsible for the communication between officers and enlisted men. They projected orders in camp and on the field. Without them, mustering and maneuvering would have been in chaos. Therefore it is important to recognize their contributions for their importance and necessity. Drummer boys served their respective causes with courage and distinction. Many made the ultimate sacrifice.

John B. Wilson of Company C in the 2nd New York wrote of when they lost their drummer during the First Battle of Bull Run:

At the third discharge a large shot came amongst our men killing two and wounding one. The ball first passed through the body of one of our drummer boys named [James] Maxwell. He gave but one sigh, and I am sure those who heard it will never forget it.

Pvt. J. D. Hicks, Company K, 125th Pennsylvania Volunteers recalled at Antietam:

Under the dark shade of a towering oak near the Dunker Church lay the lifeless form of a drummer boy, apparently not more than 17 years of age, flaxen hair and eyes of blue and form of delicate mould. As I approached him I stooped down and as I did so I perceived a bloody mark upon his forehead...It showed where the leaden messenger of death had produced the wound the caused his death. His lips were compressed, his eyes half open, a bright smile played upon his countenance. By his side lay his tenor drum, never to be tapped again.

Medal of Honor Winners

J.C. Julius Langbein was born in Germany in 1845. His family immigrated to the United States when he was still a young boy. Langbein grew up in Brooklyn, New York and at the onset of the American Civil War, he volunteered at the young age of only 15. With his parent's permission Langbein enlisted with the Union Army's 9th New York Volunteers, also known as Hawkins' Zouaves. There he served as a drummer boy. Langbein was young and small, with feminine features that earned him the nickname "Jennie" by the soldiers in his regiment. In January of 1862 his regiment joined General Ambrose Burnside's North Carolina Expedition.

During the Battle of Camden on April 19, 1862, Lieutenant Thomas L. Bartholomew was hit in the head by shrapnel and collapsed. Langbein ran to his aid despite continued heavy enemy shelling and rifle fire, and managed to guide the officer to relative safety. The regimental surgeon determined that the officer was too far gone to save but Langbein was determined that the lieutenant would not be left behind to die. He snuck him into the wagon of other wounded men headed to the federal hospital on Roanoke Island where he received life-saving care. After the Lieutenant's recovery the drummer boy was subsequently recommended for the Medal of Honor. The Medal of Honor is the highest award for bravery and valor that can be bestowed upon a member of the United States military. One such citation is that of the Medal of Honor for Johann Christoph Julius Langbein. It stated: "A drummer boy, 15 years of age, he voluntarily and under a heavy fire went to the aid of a wounded officer, procured medical assistance for him, and aided in carrying him to a place of safety."

According to his bio Langbein left the regiment in 1863 and returned to his home in New York City. He took up the uniform again in 1869, this time as an infantry officer with the New York National Guard, where he rose to the rank of captain. Returning to civilian life once again, Langbein became a lawyer and then judge in the state of New York. In 1905 he was elected commander of the Medal of Honor Legion.

Orion Perseus Howe of the 55th Illinois Infantry was just 14 years of age when he earned his own Medal of Honor for service at Vicksburg. According to the citation, "A drummer boy, 14 years of age, and severely wounded and exposed to a heavy fire from the enemy, he persistently remained upon the field of battle until he had reported to Gen. W.T. Sherman the necessity of supplying cartridges for the use of troops under command of Colonel Malmborg." Gen. Sherman himself recalled the boy's courage and tenacity. He wrote "What arrested my attention then was, and what renews my memory of the fact now is, that one so young, carrying a musket-ball wound through his leg, should have found his way to me on that fatal spot, and delivered his message..."

One of the most celebrated drummer boys in the American Civil War, William H. Horsfall, ran away from home at age 15 to serve his country in the "Great Divide". According to the Evergreen History Tour, he hitched a ride on the steamship Annie Laurie which was docked in Newport. Horsfall received the prestigious Medal of Honor for saving the life of Captain Williamson during the siege of Corinth. He was one of the youngest Kentuckians to receive this honor. The citation with his medal simply stated "Saved the life of a wounded officer lying between the lines." Horsfall served throughout the war and beyond until March of 1866 when he left the army and lived the rest of his life in Newport. He died at the age of 75.

Horsfall himself recalled his wartime experiences:

I left home without money or a warning to my parents, and in company with three other boys, stealthily boarded the steamer 'Annie Laurie,' moored at the Cincinnati wharf at Newport and billed for the Kanawha River that evening, about the 20th of December, 1861. When the bell rang for the departure of the boat, my boy companions, having a change of heart, ran ashore before the plank was hauled aboard, and wanted me to do the same. I kept in hiding until the boat was well under way and then made bold enough to venture on deck. I was accosted by the captain of the boat as to my destination, etc., and telling him the old orphan-boy story, I was treated very kindly, given something to eat, and allowed very liberal privileges.

I arrived at Cincinnati without further incident, and enlisted as a drummer boy. In the fighting before Corinth, Miss., May 21, 1862-Nelson's Brigade engaged -my position was to the right of the First Kentucky, as an independent sharpshooter. The regiment had just made a desperate charge across the ravine. Captain Williamson was wounded in the charge, and, in subsequent reversing of positions, was left between the lines. Lieutenant Hocke, approaching me, said: 'Horsfall, Captain Williamson is in a serious predicament; rescue him if possible.' So I placed my gun against a tree, and, in a stooping run, gained his side and dragged him to the stretcher bearers, who took him to the rear.

According to Deeds of Honor: Drummer Horsfall was on all the subsequent marches of his regiment. During the famous charge at Stone River he presently found himself hemmed in by rebel horsemen and hostile infantry. Even the rebels took pity on his youth and one of them shouted: "Don't shoot the damned little Yank! I want him for a cage." The plucky little drummer made a run for his life and safely got back to his regiment.

Most people are unaware that the youngest soldier ever to receive the Medal of Honor was a drummer boy named William H. "Willie" Johnston. Johnston was a drummer in Company D of the 3rd Vermont Infantry. During his service he participated in several events including the Seven Days Retreat in the Peninsula Campaign where he was said to have served in an "exemplary" fashion. During this event Johnston was the only drummer in his division to come away with his instrument during a general rout. His superiors considered this a meritorious feat, when his fellow soldiers had thrown away their guns. As a result, he received the Medal of Honor on the recommendation of his division commander, thereby becoming the youngest recipient of the highest military decoration at 13 years of age.

Johnston had enlisted at the same time as his father in June of 1861 and was assigned to a regiment that was camped outside of Washington. He was present for duty but was originally denied pay due to his age. Muster rolls from that time describe Johnston as being 11 years-old and five feet tall. His first engagement took place at Lee's Mill in Virginia on April 16, 1862. His father was shot and lost a portion of his hand while charging the enemy. Following his next campaign, the Seven Days Battles from June 25 to July 1, 1862, Johnston was cited for bravery. During a retreat many men threw away their guns and equipment to lighten the load. Johnston retained his drum and was the only drummer boy to bring his instrument off of the battlefield.

Secretary of War Edwin M. Stanton presented the Medal of Honor award to Johnston on September 16, 1863. (It is said to have been directed by President Lincoln himself although no definitive proof exists). Following the Peninsula Campaign, Johnston served as a nurse in a hospital in Baltimore and was transferred to Company H, 20th Regiment of Veteran Reserve Corps, where he played in the regimental brass band as Drum Major.

In tribute a statue honoring Johnston was erected in Santa Clarita, California. Some of his memorabilia (to include drumsticks) is on display at the Fairbanks Museum in St. Johnsbury, Vermont. A plaque was placed in his honor at Berkeley Plantation in Virginia in June, 2012 by the Vermont Civil War Hemlocks. The plaque reads: "At Harrison's Landing on July 4th, 1862, Willie Johnston — age 11, 3rd Vermont Drummer Boy played for Div. review. For keeping his drum during the arduous 7 days battles, he was awarded the Medal of Honor by Sec. of War Stanton. He remains the youngest recipient of the Medal of Honor. His gravesite is unknown. Dedicated June 2012 The Vermont Civil War Hemlocks." (Harrison's Landing is located at Berkeley Plantation in Virginia.)

Benjamin Levy was born a New Yorker and enlisted in the Union Army from Newport News, Virginia in October of 1861. He initially served with the 1st New York Infantry Regiment. While he was participating in the Battle of Glendale his drum was destroyed. Not to be deterred he took up the weapon of his injured tent mate, Jacob Turnbull, and joined in the fight. Shortly thereafter, Charley Mahorn, the color bearer, fell from a bullet wound to the chest. Levy picked up Mahorn's flag (pictured) and joined in the charge. He not only rallied the troops but protected the flag from capture. Later he was discharged but re-enlisted with the 40th New York Infantry Regiment in January of 1864. He was discharged again due to a disability in May of 1865. Levy later became the first Jewish American to be cited for and later receive the Congressional Medal of Honor. His citation reads, "This soldier, a drummer boy, took the gun of a sick comrade, went into the fight, and when the color bearers were shot down, carried the colors and saved them from capture."

Julian Scott lied about his age to enlist in the Union army. He joined Company E, 3rd Vermont Infantry at Lees Mills, Virginia on April 16, 1862. He rose from drummer boy to infantryman, and for his service he earned the Congressional Medal of Honor. Scott earned his citation for crossing a creek repeatedly under heavy fire to assist the wounded during the Battle of Lee's Mills. Following his discharge from the army Scott became an artist, who focused his painting on heroic moments of sacrifice during the war. He was especially kind to his former enemy and painted the Confederate soldier with dignity. One of his most famous paintings depicts a group of Union drummer boys in camp playing cards. After the war Scott traveled west painting Native Americans in New Mexico, Arizona and Oklahoma. Many of the paintings from these journeys hang in the University of Pennsylvania Museum of Art.

John S. Kountz earned the nickname "The Drummer-Boy of Mission Ridge" and was the subject of the poem by Kate B. Sherwood of the same name. At the age of fourteen, Kountz enlisted as a drummer boy in the 37th Ohio Infantry in September of 1861. During the Battle of Missionary Ridge the regiment's drum corps was ordered to retreat to the rear. Kountz dropped his drum in favor of a musket and was severely wounded in the first assault. He was left on the field under enemy fire until his comrades could rescue him. Kountz was shot and his right leg was shattered. It was necessary to amputate it at the hip. He remained at a hospital in Louisville, Kentucky until he was honorably discharged from the service on April 25, 1864.

He later received the Medal of Honor for his heroics on the battlefield. His citation reads "Seized a musket and joined in the charge in which he was severely wounded." Kountz remained a member of the Grand Army of the Republic since its organization in 1866 and was elected to the post of the 13th Commander-in-Chief on July 25, 1884. As one of his contributions Kountz ordered that politics be removed from the organization during the presidential election that took place during his term.

Benjamin F. Hilliker was a member of Co. A, 8th Wisconsin Volunteer Infantry. He was shot through the head near Mechanicsburg, Mississippi. Although he was terribly disfigured, Hilliker survived his wound and the war. His Medal of Honor citation reads: "For extraordinary heroism on 4 June 1863, while serving with Company A, 8th Wisconsin Infantry, in action at Mechanicsburg, Mississippi. When men were needed to oppose a superior Confederate force Musician Hilliker laid down his drum for a rifle and proceeded to the front of the skirmish line which was about 120 feet from the enemy. While on this volunteer mission and firing at the enemy he was hit in the head with a minie ball which passed through him. An order was given to "lay him in the shade; he won't last long." He recovered from this wound being left with an ugly scar.

40 Musicians have been awarded the Congressional Medal of Honor since its inception.

Memorable Drummer Boys

Some young musicians marched in the ranks of the U.S. Colored Troops while contributing in their own way. Unlike their counterparts in the South, blacks, both free and ex-slave were looked upon as soldiers and not camp servants. Grateful for their newfound freedom many Southern slaves savored the opportunity to line up in the Union ranks and raise their muskets toward their former oppressors. Free men from the North took the opportunity to serve as their brother's keeper. Throughout the war drummer boys provided essential camp and field communications.

One African-American drummer boy of particularly noteworthy service was A.H. Johnson. At the age of 16, Alexander H. Johnson was the first African American musician to enlist in U.S. military, joining the 54th Massachusetts Volunteers under Robert Gould Shaw. Johnson was adopted by William Henry Johnson, the second black lawyer in the United States and close associate of Frederick Douglass. After the war Johnson told an interviewer that he had "beat a drum every day he has been able since childhood."

According to an article titled *Alexander H. Johnson: The first drummer boy* (by Meserette Kentake) Johnson quickly established himself as a talented drummer as he and the rest of the rank and file learned the art of soldiering. He was with the unit when it left Boston for James Island, S.C., where it fought its first battle.

The skirmish, along the South Carolina coast near Charleston, occurred on July 16, 1863. Johnson noted, *"We fought from 7 in the morning to 4:30 in the afternoon, and we succeeded in driving the enemy back. After the battle we got a paper saying that if Fort Wagner was charged within a week it would be taken."*

Two days later the 54th unsuccessfully stormed Confederate-held Fort Wagner on Morris Island while sustaining massive casualties. Johnson recounted, *"Most of the way we were singing, Col. Shaw and I marching at the head of the regiment. It was getting dark when we crossed the bridge to Morris Island. It was about 6:30 o'clock when we got there. Col. Shaw ordered me to take a message back to the quartermaster at the wharf, who had charge of the commissary. I took the letter by the first boat, as ordered, and when I returned I found the regiment lying down, waiting for orders to charge. The order to charge was given at 7:30 o'clock."*

Johnson remained in the 54th until the end of the war. In the summer of 1865 he returned to Massachusetts, bringing the drum that he carried at Fort Wagner with him. Four years later he married, settled in Worcester, Mass., and organized "Johnson's Drum Corps." He led the band as drum major, and styled himself "The Major."

In 1897, a memorial to the 54th sculpted by the artist Augustus Saint-Gaudens was unveiled in Boston. The bronze relief depicts Colonel Shaw and his men leaving Boston for the South with a young drummer in the lead — a scene reminiscent of the July day in 1863 when Shaw and Johnson marched at the head of the 54th to its destiny at Fort Wagner. In 1904, Johnson visited the monument during an event hosted by the Grand Army of the Republic, the influential association of Union veterans. Many of those in attendance pointed out the resemblance of the young lead drummer and it is said that Johnson felt a great sense of pride for his participation in the war. Today the statute remains as a timeless tribute to both Johnson and the men he served.

Perhaps the most photographed drummer boy of the American Civil War, Robert Henry Hendershot, was known as the "Drummer Boy of the Rappahannock." His nickname supposedly came from his reputed heroics at the Battle of Fredericksburg, Virginia, in December of 1862. Hendershot enlisted in Company B, 9th Michigan Infantry in March of 1862, and was taken prisoner that July at the Battle of Stones River. After his release, he joined the 8th Michigan Infantry, although he suffered from regular seizures.

While awaiting discharge for epilepsy, Hendershot arrived on the banks facing Fredericksburg where the Army of the Potomac was preparing to attack the city. The Army of Northern Virginia was waiting on the banks of the Rappahannock River, defending the city while pontoon bridges were being built. The delay enabled General Robert E. Lee to move the Confederate army into a formidable position. When the Union engineers arrived, they came under attack from rebel sharpshooters, so on December 11, 1862 the 7th Michigan Infantry volunteered to cross the river under enemy fire and drive the rebel sharpshooters from their nests. According to an account of the events:

[Hendershot's wanderings had taken him to the riverbank that morning. He later claimed that he helped push off the first boat, slipped when he tried to climb aboard, and made the voyage across the river while clinging to the gunwale. A dispatch from the scene describes "a drummer boy, only 13 years-old, who volunteered and went over in the first boat, and returned laden with

curiosities picked up while there." A correspondent for the Detroit Advertiser and Tribune wrote that the boy belonged to the 8th Michigan Infantry. Reports of the episode appeared in the press.

The young hero remained nameless until late December, when Hendershot visited the offices of the Detroit Free Press and Detroit Advertiser and Tribune, claiming to be the "Drummer Boy of the Rappahannock." Hendershot's story was repeated in national papers, including the New-York Tribune. Its publisher, Horace Greeley, presented Hendershot with a silver drum. For the next eight weeks Hendershot performed at the P. T. Barnum museum, and then spent a few weeks more in Poughkeepsie, New York, at the Eastman Business College, which had rewarded his heroism with a scholarship.]

Many historians have questioned the story of the "Drummer Boy of the Rappahannock." The only ones who knew the truth were the witnesses who were present at the boat's launching and Hendershot himself. Following the war in July 1891 Hendershot posted a letter to the Grand Army of the Republic (GAR) newspaper, the National Tribune, restating his claim to the title "Drummer Boy of the Rappahannock," as well as that of "youngest soldier." He was by then one of the best known veteran drummer boys in the country. Despite the ongoing controversy Hendershot always stood by his claims before dying of pneumonia on December 26, 1925.

Riding on the coattails of his fame Hendershot often made public appearances. In 1891 The Grand Army of the Republic held their national meeting in the City of Detroit. Hendershot was invited to participate in the parade. Members of the 7th Michigan who had contested Hendershot's story were livid at the thought of who they considered to be a fraud participating in their affair. Hendershot performed "Marching Through Georgia." for a crowd of 200 with a new drum that was gifted to him. His son performed at his side on the fife. A local dignitary recited the popular poem "The Drummer Boy of the Rappahannock" much to the crowds cheers.

The opening stanza went:

'T WAS a question if the nation should such tender youth employ
As Robert Henry Hendershot, the little drummer boy
A prodigy at drumming—being only twelve years old—
And a prodigy of valor as the story has been told
At Fredericksburg's great battle
The soldiers heard the rattle
Of his drum

Offended members of the 7th Michigan publicly challenged the celebrity to produce even one witness who had seen him at Fredericksburg. In response Hendershot produced letters from President Lincoln, General Ulysses Grant, and Horace Greeley attesting to his heroism. The crowd was nearing the point of violence when the 7th's own drummer boy, John S. Spillane (a captain of the Detroit police force), entered the room. "There is the drummer boy of the Rappahannock!" shouted a fellow veteran. They hurried Spillane to the platform and unceremoniously removed Hendershot from the room. After the meeting, the citizens of Detroit presented Spillane with a medal proclaiming him, not Hendershot, to be the real "Drummer Boy of the Rappahannock."

The youngest drummer boy, and perhaps the youngest recruit the U.S. Army has ever seen was Edward Black. On July 4, 1862 the then 8 year-old drummer boy was recruited as a member of the 21st Indiana Infantry. He was told to return home but quickly disobeyed that order. Later that year, and one year older, he enlisted again but this time with his father's permission. He was captured at the Battle of Baton Rouge, but released when the city fell. Black was discharged in September of 1862. He once again re-enlisted in February of 1863 and served with the 1st Indiana Heavy Artillery until January of 1866. He died at age 19 in 1872 and was buried along with his twin brother Edwin in the Crown Hill National Cemetery in Indianapolis. His drum is part of the collection at the Children's Museum of Indianapolis.

"Little Morris" was one of the youngest boys enlisted on either side during the Civil War. Morris survived the conflict and left behind a legacy that became a legendary story in the press. At the tender age of 10 ½ Morris was presented to Captain J. Murray of Company D., 19th Virginia Infantry. Despite his young age he was accepted to fulfill the role of a drummer boy. At the time he was the youngest enlistee on either side of the war. After being treated harshly by the men of Company D. Morris transferred to Company A., 19th Virginia Battery of Artillery where he served under Captain J.F. Chalmers until the end of the war. Following the war Morris grew up to be a popular citizen around Richmond who shared his experiences from the war. His drumsticks and portrait were put on display.

The youngest boy to die in the Civil War also happened to be a drummer boy. Charles Edwin King was 12 ½ years-old when he enlisted in the Union Army on September 12, 1861. His father had great pains about his son going off to war but Capt. Benjamin Sweeney of Company F, 49th Regiment, Pennsylvania Volunteers reassured the nervous parent that he would try and keep Charles out of harm's way. The young drummer saw his first action at the Battle of Williamsburg where the Union Army retreated off of the Virginia Peninsula after engaging Robert E Lee's Army of Northern Virginia. Prior to the next battle Charles was promoted to Drum Major and made a full Union soldier. He was only 13 years-old.

In September of 1862 the regiment marched to Western Maryland to take part in the Battle of Antietam which took place on September 17. The 49th Regiment took few casualties but things turned when a Confederate shell exploded at the rear of the line. Several men were wounded, including Charles, who was shot by shrapnel. The wound was through the body and deemed mortal. His father, who was also serving in the Union Army, was summoned to the field hospital. He did not make it in time. After three days of pain Charles died on

September 20. No one knows what happened to his remains. Some have guessed that he is buried at Antietam but there is no record that supports that theory. A monument was placed near his parent's graves in the Green Mount Cemetery in Westtown to preserve his sacrifice to preserve the Union.

Clarence MacKenzie was a mere 12 years-old when he marched off to war as a member of Brooklyn's 13th Regiment. Tragedy befell him in June of 1861 while he was encamped at Annapolis, Maryland. It was during a training drill that MacKenzie was accidentally struck by a stray ball fired by his fellow soldiers. As a result he became Brooklyn's first casualty of the Civil War. MacKenzie's body was returned home and buried in a public lot on the Hill of Graves at Green-Wood cemetery. He was later relocated to the Soldiers' Lot which Green-Wood donated specifically for Civil War Veterans. His grave is marked with a striking white bronze monument forged in his likeness. The ornate pedestal carrying the statue stands approximately ten feet in height and is inscribed: "ERECTED BY THE DRUM AND BUGLE CORPS OF THE 13TH REGT. N.G., S.N.Y., IN MEMORY OF CLARENCE D. MACKENZIE, BORN FEB. 8, 1849, DIED AT ANNAPOLIS, MD., JUNE 11, 1861, AGED 12 YRS, 4 MOS, 3 DYS."

At the age of 10 Johnny Clem tried to enlist in the newly formed 3rd Ohio Regiment but was turned away again and again. He later attempted to join the 22nd Michigan who eventually allowed him to follow the regiment as a mascot and drummer boy. Clem became a national celebrity after his actions at Chickamauga.

Armed with a musket that was sawed down to accommodate his size Clem joined in the defense of Horseshoe Ridge on the afternoon of September 20. As Confederate forces breached the line and surrounded the defenders a colonel was said to have called him a "damned little Yankee." Rather than surrender Clem shot the officer and made his way back to the Union lines. For his brave actions Clem was promoted to a sergeant and the youngest noncommissioned officer in the Union Army.

According to his bio: In October 1863, Clem was captured in Georgia by Confederate cavalrymen while detailed as a train guard. The Confederates confiscated his U.S. uniform which reportedly upset him terribly—including his cap which had three bullet holes in it. He was included in a prisoner exchange a short time later, but the Confederate newspapers used his age and celebrity status for propaganda purposes, to show "what sore straits the Yankees are driven, when they have to send their babies out to fight us."

Clem fought in several battles including Perryville, Murfreesboro, Kennesaw and Atlanta. He was wounded twice and was discharged from the army at the age of 13. He would later attend West Point. Clem died in 1937 and is buried at Arlington National Cemetery.

The topic of Black Confederates continues to be one of the most controversial subjects in Civil War scholarship. Throughout the course of the war, Confederate officers routinely brought their slaves with them to act as camp servants and mess cooks. This was done as both a reflection of the officers' social status and for the domestic services provided by the slaves. In some cases, these African Americans would be issued uniforms, and their typical responsibilities included cooking, washing clothes and cleaning quarters. In addition, those slaves with a musical talent were often called upon to sing, dance and play tunes to entertain their masters' staff or messmates. The sincere nature of these relationships is required to be judged on an individual basis, but it is fair to say that the overwhelming majority of blacks in Confederate camps were acting in the role of servants rather than soldiers. This topic has been aggressively debated to this very day. Historians routinely differ with those who have propagated what they consider to be a myth. Newfound information continues to support their theory of mythical historical memory.

One African American who is believed to have enlisted in the Confederate army and served as a free man was Henry "Dad" Brown, a Confederate drummer from Darlington South Carolina. Brown was a veteran of the Mexican, Civil, and Spanish-American Wars. He is said to either have been born free or as a slave that was able to purchase his freedom. According to his roadside marker he joined the Confederate army in May of 1861 as a drummer in the Darlington Grays, Co. F, 8th S.C. Infantry. After they disbanded he enlisted as a drummer in Co. H, 21st S.C. Infantry in July 1861 and served in that outfit for the rest of the war. Years later he was made a member of the Darlington Guards where he held a membership from 1878 until his death in 1907.

It was said that Brown "captured" a pair of Yankee drumsticks at the Battle of Second Manassas. (They are on display at the Darlington County Historical Society museum.) Confederate Gen. W.E. James recalled Brown's valor in a written account of that battle:

...on the 21st of July '61 the regiment was stationed at Mitchel's Ford on the South side of Bull Run. The battle began two miles above and at 12 o'clock the regiment was ordered to go where the battle was raging. As soon as the order came Henry began to beat the long roll. This indicated to a battery on the other side of the Run the position of the regiment and the shells began to fall thick and fast. It was some time before the Colonel could stop him but he was

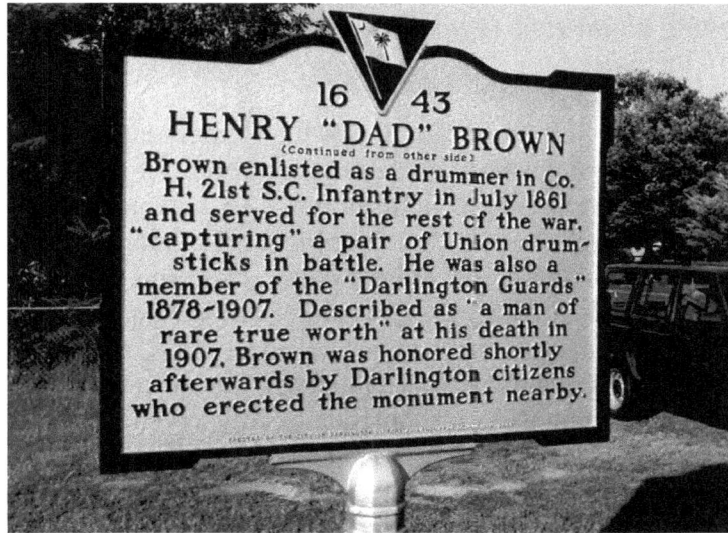

beating all the time regardless of the danger. He followed on to the battlefield and was under fire with the others.

Brown's service and dedication to the Confederate cause has been debated for decades. That argument has not affected his opportunity for commemoration. In 1907 a grave and monument were erected in his honor. The citizens of Darlington were said to have referred to him as "a man of worth." A spiraling 20-foot marble obelisk was erected in his honor by both black and white members of the community. In 1990 his monument underwent restoration and was rededicated with a 21-gun salute. The ceremony was attended by a crowd of 200 people.

Thomas Jefferson Cole was a musician in Company E of the 1st Regiment, South Carolina Volunteer Infantry. This Confederate unit served in many sizes and capacities throughout the war and was present at the bombardment of Fort Sumter. After the war started they became part of the Provisional Army of the Confederate States of the Army of Northern Virginia. They participated in many of the war's major battles to include: Second Manassas, South Mountain, Sharpsburg, Fredericksburg, Gettysburg, The Wilderness, Spotsylvania, Cold Harbor and they were present at the South's surrender at Appomattox Court House. Unfortunately little is known about young T.J. Cole other than his photograph and service record. Still they offer a glimpse into the service of a drummer boy during America's great conflict. Cole lived to be 75 years-old.

Albert Henry Woolson was the last surviving Civil War veteran on either side whose status is undisputed. Woolson's father had enlisted in the Union Army but was wounded at the Battle of Shiloh. He later died of his wounds at a hospital in his home state of Minnesota. Following in his father's footsteps Woolson enlisted as a drummer boy in Company C, 1st Minnesota Heavy Artillery Regiment. Thankfully the company never saw action and he was discharged in September of 1865.

Following the war Woolson returned to Minnesota where he became a member of the veteran's organization known the Grand Army of the Republic. At the age of 99 he and fellow drummer-boy Frank Mayer marched together in the 1949 New York City Memorial Day Parade where they laid a wreath at the tomb of General Ulysses S. Grant. In 1953 Woolson became the senior vice commander in chief of the G.A.R. where he informally served for the remainder of his life.

Woolson died in Duluth in August of 1956, at what was then thought to be the age of 109. He was buried with full military honors by the National Guard. President Dwight D. Eisenhower paid tribute to Woolson saying that "The American people have lost the last personal link with the Union Army." In 1956 a monument of Woolson was erected at the Gettysburg National Military Park as a memorial to the Grand Army of the Republic.

Anna Glud of Oakland, California represented at least one Civil War drummer girl. Eager to serve her country she donned a Federal uniform and went off to war. Using the name "Tom Hunley" she served in the Federal Army for two years. She claims that General U.S. Grant found out about her deception. Grant was inspecting Tom's regiment and upon seeing the diminutive drummer boy, ordered him mustered out as too young for service. Anna's father pleaded with Grant, telling him that his daughter was motherless and that he did not want to leave her home alone. The general swore himself to secrecy and canceled his order to muster "him" out. Anna later wrote:

During all the time that I was in the Army many remarked that I looked more like a girl than a boy. But not one soldier actually found it out. Father and I kept so constantly together that I was always protected. Had I not had his assistance at all times, I doubt that I could have stood the rigors of a soldier's life. Why, in a battle near Davisville, where 7,000 Confederates and Northerners were killed, our little body of men literally had to climb over the bodies of dead soldiers in order to fight our way out. My little feet were red with blood. And when we were mustered out in the fall of 1864 there were but 17 members of our company left.

After the war Anna still disguised herself as Tom and her father attempted to settle down on a farm in Indiana. The war was apparently difficult on her father. He died six months later. His wife and four sons had already died. At this time, "Tom" went back to being Anna once again.

Chauncey H. Cooke of the 25th Wisconsin also deceived his superiors. He wrote of his mustering experience:

Every one he suspicioned of being under 18 he would ask his age. He turned out a lot of them who were not quite 18...Seeing how it was working with the rest, I did not know what to do...I saw our Chaplain and he told me to tell the truth, that I was a little past 16, and he tho't when the mustering officer saw my whiskers he would not ask my age. That is what the boys all told me but I was afraid. I had about made up my mind to tell him I was going on 19 years, but thank heaven I did not have a chance to lie. He did not ask my age. I am all right ... but the sweat was running down my legs into my boots when that fellow came down the line and I was looking hard at the ground fifteen paces in front.

One of the last surviving Confederate drummer boys was Martin Delphos Luther. He enlisted on January 1, 1864 at Buncombe County, NC as a Private into Co. I, NC 25th Infantry. Martin participated in the battles of Suffolk, VA., and Plymouth, N.C. At Plymouth, he was seriously wounded and left for dead on the battlefield. He was later taken to a field hospital and in 90 days he was well enough to rejoin his regiment in the Overland Campaign in Virginia.

During the siege of Petersburg, the company's drummer was killed and Luther took his place, serving in that capacity until the end of the war. He is believed to be one of the few remaining drummers in General Robert E. Lee's Army of Northern Virginia before the surrender at Appomattox Court House in April of 1865.

In 1925 Luther was invited to speak to Union Veterans at their annual Grand Army of the Republic encampment in Athens, Tennessee. Luther told them that as the years passed the veterans of the GAR were as near to him as if they had worn his color during the war. He died at Athens in southeast Tennessee on September 22, 1925.

Henry Monroe was a member of the famous U.S. Color Troops 54th Massachusetts Voluntary Infantry Regiment. He was just 13 years-old when he drummed by his commander's side during the disastrous assault on Fort Wagner on July 18, 1863. The 54th suffered massive casualties including having their commander Col. Robert Gould Shaw killed as he led his men forward. This battle inspired the film Glory.

Later in his life Monroe described the attack as a 'slumbering volcano' that 'awoke to action and poured forth sheets of flame from ten thousand rebel fires, and earth and heaven shook with the roar of a hundred pieces of artillery.'

After the war Monroe attended public schools in Boston and graduated at the head of his class. He was the only African-American to attend his courses. He went on to teach in the Freedman's Bureau and President Ulysses S. Grant appointed him an inspector of customs at the Port of Baltimore. He also went on to publish a newspaper.

Charles F. Everett was a drummer for the 57th Massachusetts Infantry Regiment and was from Worcester. He is believed to have been between the ages of 12 and 14 when he was killed at the Battle of the Wilderness in Virginia, May 5-7, 1864. That battle was particularly horrific where 162,000 Union and Confederate troops fought to a stalemate, with nearly 29,000 casualties. The flash from the weapons caused brush fires to erupt. The wounded couldn't get out of the woods and were literally burned to death. According to reports Everett was told by his commanding officer to stay back with the horses and wagons.

Like many of his peers in this book, he got a gun, a satchel of ammunition, and went ahead with the men. Soon after he was mortally wounded in the hip. Captain Warren Galucia was the commander of Company E of the 56th and a friend of Everett's family. He ordered Sergeant Robert Horrigan to make his way to the front to check on the boy's predicament. Upon reaching the front he found the drummer severely wounded. He left him lying at the base of a tree. His body was never found. After he was killed in action his mother received a pension of $8 a month. His name is one of 398 names on the Soldiers and Sailors Memorial, the Civil War monument on Worcester Common that commemorates the local men who lost their lives in the 1861-1865 conflict.

Julius W. Pell enlisted as a drummer boy in Company B, 11th West Virginia Volunteer Infantry at Burning Springs, Wirt County, on December 24, 1861. He was only a little over 11 years-old at the time. He served in that capacity for over three years. When the war started Pell's father W.F. Pell was already a Colonel of the local militia. His community was mostly comprised of men with Southern sympathies and Mr. Pell was asked to lead a company in the field to fight for the Confederacy. A strong proponent of preserving the Union he refused the invitation. He enlisted as a private in Company A, 11th W. Va. Volunteer Infantry. Due to his prewar rank he was immediately promoted to a Captain of Company B.

His two young sons enlisted following their father's example. Julius served under Old Glory as a drummer throughout the war. His service for the most part was in the armies of the Shenandoah and Potomac. Pell was honorably discharged at Camp Deep Bottom, Virginia, on January 4, 1865. His real age at the time of his enlistment was not given but he was said to be "the most diminutive drummer boy in the service." He claimed to be the youngest recruit at the time of his service. Another merit claimed by Pell is that he was able to cast a vote for Abraham Lincoln for President of the United States while a fraction over fourteen years of age. After he was discharged from the army he was not permitted to vote until he reached the legal age.

William T. Simpson served as a drummer boy with the 28th Regiment, Pennsylvania Volunteer Infantry, Company A (below). He was promoted to Principal Musician of the Regiment. William survived the war and attended many veteran ceremonies. He died in 1940 at the age of 92.

Sometimes the drummer boy's contributions were not appreciated by their older companions. Drumming meant marching, and lining up, and moving into battle. All of which were detested by the common soldier. Although the drummers were not to blame they were often the source of the soldier's frustration. On the other hand, it was difficult to ignore the courage at which the drummer boys performed their duties. Some of the other drummers remembered today include:

Edward Black, 21st Iowa Regiment

Thomas Camp, 11th Wisconsin, Company F, Harvey Zouaves

Charley Common, 52nd Ohio Regiment

Lyston D. Howe, 15th Illinois Volunteers

Jackson, 79th U. S. Colored Troops, Louisiana

Charles Monell, 165th New York, 2nd Duryee Zouaves

John M. Raymond, 11th Michigan Volunteer Infantry

William C. Richardson, 104th Ohio Regiment

Franklin Searis, Company E, 19th Wisconsin Volunteers

Johnnie Walker, 22nd Wisconsin Regiment

Letters Home

Drummer boys, like any soldier, wrote home to their family to reassure their family members that they were still alive and faring relatively well. This was often done to alleviate the worry of their loved ones. After all, as volunteers they did not have to be in their current predicament. Here are some letters as examples of the kinds of correspondence that was written from home and on campaign.

Transcript of one of the letters of Felix Voltz, a drummer boy in the 187th New York Volunteer Regiment during the Civil War.

According to author J. Arthur Moore's bio on Voltz: "Felix Voltz ran away from home on January 30, 1865 to enlist (to his family's dismay). He mustered out with the company on July 1, 1865, at Arlington Heights, Virginia, and served as a drummer in the 187th Regiment, New York Volunteer Infantry for five months. Felix wrote letters to his family in Elmira, New York, which described the rigors of Union Army life from February through June 1865."

One thing in particular is the revelation that many of the drummers were responsible for furnishing their own drums as backing was directed towards more pressing wartime necessities. As the vast majority of drummer boys were young they were dependent on their parents or guardians to provide the means in which to obtain the instrument. We can assume their fears were subsided knowing that their boys were not assigned to a combat role and were serving as musicians. (Note: This letter is held in the Special Collections Department of the University Libraries at Virginia Tech):

March 3d / 65 187 Regt

Dear Parents Brths & Sisters

Now I will let you know how I got in the Drum Chor I had to go on Picket Duty the other day and when I came back I got sick for two or three days but I got over that and then I went to Tony the Orderly and ask him if they had A Drummer for our Company Says he No sir then he told me to wait A day or two and he would set about it then he to Drum Major and when he come back he told Me to go over to the Drum Major he wanted too see Me when I come over there who was Drum Major was Joe Koack and he told Me if it was possible he would get Me in and then he came over and told me to give up my Musket and come with him then he said he would try and see if he could get Drum for Me here but he said I could not draw any government Drum down here he told Me to write Home fore one and have it send here you Tony can go and do this favor for Me he said the best and cheapest place you can buy one is on the corner of Main and Tiagarer Sts a new music Store and please buy a good one and I will make it all right as soon as I get My Bounty and he Joe told Me best way and the quickest way to send it would be by Mail / they tell us we will get our Bounty the 15th of this Month then I will send home all I possible can. No More news this time I will write again as soon as possible please tell Mother not to wearry herself about Me for I am allright yet and I hope will be so for the next year and tell here I am in no danger what so ever all I have to do is to take care of Me and my Drum and learn how to Drum as soon as possible…(follow up letter)…my Drum arrived here yesterday alright in good Order and Joe Roach says that you could not send A better one for here in the Army I thank you Brth A W for doing that favor…

I remain your truly Son and Brother.
Felix Voltz

Recollection written by Justin S. Keeler, a drummer boy, 17th Connecticut:

I was a drummer boy in my regiment and of the ten sheepskin pounders as we were called. I was the only one to bring his "music box" away when the stampede began. I prized it very highly, as the boys in my company presented it to me when the regiment was stationed on the heights in the suburbs of Baltimore. When the doctor caught sight of me he said, "Young man, throw that drum down and catch hold of this stretcher; I want this captain taken down to the hospital." In vain I pleaded for my old drum, as I wanted to show the boys when we again reached camp, that my love for their gift was not forgotten "under the most trying circumstances." The doctor however, was inexorable; the captain's life was of more consequence than the drum so the latter had to be sacrificed. I took it off and sorrowfully stood it against a tall pine tree, covering it with leaves and branches, as I thought I might return for it if our army held its present position.

Drummer boy Harry Martyn Kieffer of the 150th PA Vols. wrote to his father:

Near Rappahannock River, April 28

Here is an old army letter lying before me, written on my drum-head in lead pencil, in that stretch of meadow by the river, where I heard my first shell scream and shriek: – Dear father – We have moved to the river, and are just going into battle. I am well and so are the boys. – Your affect son. "Harry"

He later wrote about his experience as a stretcher bearer at Gettysburg:

"[I am called] away for a moment to look after some poor fellow whose arm is off at the shoulder, and it was just time I got away, too, for immediately a shell plunges into the sod where I had been sitting, tearing my stretcher to tatters."

A 16 year-old drummer, John A. Cockerill, who was at Shiloh, later wrote:

"I passed... the corpse of a beautiful boy in gray who lay with his blond curls scattered about his face and his hand folded peacefully across his breast. He was clad in a bright and neat uniform, well garnished with gold, which seemed to tell the story of a loving mother and sisters who had sent their household pet to the field of war. His neat little hat lying beside him bore the number of a Georgia regiment... He was about my age... At the sight of the poor boy's corpse, I burst into a regular boo hoo and started on."

Here is a letter home from an unknown author:

WASHINGTON, May 1861,
with the (NY) Twelfth Regiment

A cold, rainy and uncomfortable day – the first since the Twelfth regiment entered Washington – has temporarily suspended parades and outside drills, and in the interim this afforded I have coveted a brief space in your columns to communicate with the thousands of New York city who feel a deep interest in the movements of this regiment.

...One feature connected with this regiment is its band and drum corps. They are both under the charge of Drum Major Smith, an accomplished musician and a thorough tactician, and have won golden opinions from every one by their excellent music and their proficiency in drill.

Mother Calista Hubbard to her 14 year-old son and drummer boy Lucien Welles Hubbard:

Aunt Sarah tells me that when she was over to the Hospital that Fred Standish said that the drummer boys did not have to be much exposed in time of battle unless you are a mind to, but he says you and George Allen are always around.

Now Lucien if you love your mother for my sake do not expose your life unnecessarily, for we are commanded to use all means to preserve our lives and the lives of others. Remember that you have a precious soul which must spend an eternity in happiness or misery. You can't tell how much I feel on your account surrounded by thieves and every thing that is bad and evil examples of every kind. Don't you sometimes think you would like to be at home and have a quiet home life

once more? I sometimes fear that I shall never see you again on earth. But if not I hope we may meet in Heaven.

Last letter written by 15 year-old drummer boy John Ross Wallar as he lie dying in a field hospital:

Dear Sister father Mother and friends

I recievd your letter But I don't think I Ever shall see another that you write this is Friday night But I don't think I will Live to See Morning But My Kind friends I am a Soldier of Christ I will Meet you all in Heven My Leg Has Bin taking of above My nee I am Dying at this time so don't Morn after Me fore I Have Bleed and died fore My Country May God Help you all to pray fore Me I want you all to Meet Me in Heven above Dear Sister you wanted to Know if My Leg would be Stiff God Bless Your Soul Sister I will be Stiff all over be four twenty four ours My wound Dresser is writing this Letter fore Me when you get this Letter write to Alexander Nelan fore I wont Live till Morning so good By My friends May God be with you all good by God Bless My poor Soul

Hamilton Wetherby from Company C, 111 Regiment, New York State Volunteers wrote to his sister of his first experience in battle:

...You wanted to know whether I was in the battle or not. I made out to be. I do not think that I should go to bed and let them fight for me. I could not do that, I tell you. It sounded funny to hear the bullets fly all around my head. They come right close to my head but did not hit me but over reckoned. There was a lull, just scaled Lt. (the Lieutenant's) nose and knocked off a little piece of skin. That was a pretty close call for his nose. Tell Emma I will answer her letter in a day or two. Must close now, so good bye from, Your affectionate, Hamilton. I send my love to all the girls in Victory.

I must close now so good by from your Brother, Hamilton Wetherby.

Wetherby served as a drummer boy but was later promoted to a private. He was killed in action during the Battle of the Wilderness in Virginia. He was first buried at Cook's farm in Spotsylvania County and then re-interred and buried in the national cemetery in the nearby City of Fredericksburg.

Drummer Billings P. Sibley of Mankato, Minnesota wrote home of a burial of one of his comrades:

Fort Snelling
Sept. 1861

Dear Mother,

There was one of our boys died last Saturday and was buried Sunday with all the military honors. There was 24 guns fired over his grave; and had all companies (there were 4) and a band of music in front and tunes very appropriate for the occasion. And when the procession had left the ground, the father of the boy arrived and had to dig the corpse up again.

It is very near time to retire. Therefore I must bid you goodbye.

B.P. Sibley

The trauma experienced by Civil War drummer boys no doubt left them troubled after the conclusion of the war. They had served alongside adults with the same courage and distinction. Some became prisoners, some were killed and others died of disease. Today in order to properly honor these young soldiers we must remember the nightmarish conditions in which they performed their tasks. Their playing was a major contribution to the army, no matter which cause, both on the field and off.

At Gettysburg National Cemetery the Drummer Boy's contribution is forever preserved on their graveyard marker:

The Muffled Drums Sad Roll Has Beat
The Soldiers Last Tattoo
No More Life's Parade Shall Meet
The Brave and Fallen Few

The Life of a Drummer Boy

698/27

War Department,

RECORD AND PENSION DIVISION,

Washington, **AUG 9 1889** *, 188 .*

Respectfully returned to the Commissioner of Pensions.

John Sissons, a prio. rec't of Company D, 51 Regiment Ohio Volunteers, was enrolled on the 27 day of Feb, 1864, at Coshocton, for 3 years, and is reported: on Roll Mch & Apr. '64 present. Same July & Aug '64 (must on file.) Sept & Oct '64 not on file. Rolls Co. D. to which transferred Nov & Dec. '64 to Feb 28 '65 present daily duty as musician in Reg't Band. Mch & Apr. '64 absent on daily as musician in Reg. Band, To Aug 31 '65 present on daily duty as drummer. M.O. with Co. at Victoria, Tex, Oct. 3 1865. Returns May, June, Sept & Oct '64 do not report him absent. No further information.

Some drummer boys earned pensions after the war. John Sisson played flute in the regimental band and was a drummer in the 51st Ohio, Co. D.

Civil War drums were made primarily in the important industrialized centers of the Northeast: Boston, New York and Philadelphia. An American bald eagle most commonly emblazoned the Federal Army drums but sometimes the Confederates used it as well. Federal drums were also decorated with 13 stars for each of their 13 states. Confederate states were represented with 11 stars.

The romance and innocence of the drummer boy often made him an interesting subject matter for battlefield artists and painters. Drummers were often portrayed in both paintings and in the newspapers. The great American artist Winslow Homer, who had covered the war as a sketch artist, placed a drummer in his classic painting *Drum and Bugle Corps*.

Drummer boys often returned to veteran reunions to perform once again with the regimental bands that they had served with during the war. Many had become drum instructors or drum majors.

Some boys would put a note with the number 18 in their shoes when applying for the army. This way they could say "I'm over 18" without really lying. Some historians estimate that as many as 20% of the soldiers who fought in the Civil War were under the age of 18.

Regulation drumsticks were usually made from Rosewood and were 16"-17" in length. Most drummers today use 15".

"The Drummer Boy of Shiloh."
The song lyrics go:

On Shiloh's dark and bloody ground,
The dead and wounded lay,
Amongst them was a drummer boy,
Who beat the drum that day.
A wounded soldier held him up,
His drum was by his side;
He clasped his hands then raised his eyes
And prayed before he died.
He clasped his hands then raised his eyes
And prayed before he died.

This photo depicts Company A., 5th Regiment of Georgia Volunteer Infantry aka "The Clinch Rifles of Augusta" the day before they mustered out. A.K. Clark, the boy in the center with the drum, fortunately preserved a copy of the picture. Fifty years later he wrote: "I weighed only ninety-five pounds, and was so small that they would only take me as a drummer. Of the seventeen men in the picture, I am the only one living." As the war progressed Clark grew into a young man and became a real soldier.

Kepi hats often had to be specially made for drummer boys due to their small size. This hat belonged to a Union drummer boy.

Next Page: Thomas Nast illustration (1863)

THE DRUMMER.

HIS TOILET — THE FAVORITE — IN CAMP — HIS DAILY BREAD

BOY OF OUR

OFF TO THE WAR. — HOME AGAIN!

REGIMENT.

WRITING HOME. — IN ACTION. — NEWS FROM HOME.

THE DRUMMER BOY OF OUR REGIMENT—EIGHT WAR SCENES.

Forever Young

Drummers were often photographed in studios prior to and after the war. These portrait cards were given to family members as keepsakes to remind them of their loved ones.

Wounded Drummers

Here are some drummer boys who made a sacrifice.

Jimmy Doyle was wounded
at the Battle of Chickamauga.

Edward (William) Black.
At 12 years-old, his left hand
and arm were shattered.

Unknown wounded drummer
boy with both arms amputated.

Orion Perseus Howe had a
bullet rip open his thigh during
the assault on Vicksburg.

Portrayed In the Press

Drummer boys were often romanticized in the papers and they were often portrayed at ease as opposed to suffering at war.

Reveille in Camp (Harper's Weekly, July 1863)

Teaching the Negro Recruits the Use of The Mine Rifle (Harper's Weekly, March 1863)

1867 Civil War book *The Boys in Blue*

Portrayed in Prose

The story of the Drummer Boy has been captured in poem form

He was just a boy of sixteen

When he joined the Union army.

He had a drum and a pair of sticks

And a dream of glory and victory.

He marched along with the soldiers,

Keeping time with his steady beat.

He saw the smoke and the fire

And the blood on the fields of Gettysburg.

Now he haunts the battlefield,

Playing his drum for eternity.

He's the ghost of the drummer boy;

The lonely spirit of Gettysburg.

He never fired a shot or swung a sword,

But he gave his life for his cause.

He was caught in the crossfire

On the third and final day.

He fell down with his drum beside him

And his sticks still in his hands.

He never got to see his home again

Or the end of the war he fought in,

Now he haunts the battlefield,

Playing his drum for eternity.

He's the ghost of the drummer boy;

The lonely spirit of Gettysburg.

Some say they hear him at night,

When the moon is full and bright.

They hear his drum echoing in the air;

They feel his presence everywhere.

He's still looking for his comrades.

He's still waiting for his orders.

He's still hoping for some peace.

He's still trapped in his memories.

Now he haunts the battlefield,

Playing his drum for eternity.

He's the ghost of the drummer boy;

The lonely spirit of Gettysburg.

He's the ghost of the drummer boy;

The saddest song of Gettysburg.

By Matthew R. Callies

Portrayed in Song

The following is William "Shakespeare" Hays' popular 1862 song "The Drummer Boy of Shiloh":

"Look down upon the battlefield,
Oh Thou, Our Heavenly Friend,
Have mercy on our sinful souls."
The soldiers cried, "Amen."
There gathered 'round a little group,
Each brave man knelt and cried
They listened to the drummer boy,
Who prayed before he died.

"Oh, Mother," said the dying boy,
"Look down from heaven on me.
Receive me to thy fond embrace,
Oh, take me home to thee.
I've loved my country as my God.
To serve them both I've tried"
He smiled, shook hands —death seized the boy,
Who prayed before he died.

Each soldier wept then like a child.
Stout hearts were they and brave.
They wrapped him in his country's flag
And laid him in the grave.
They placed by him the Bible,
A rededicated guide
To those that mourn the drummer boy
Who prayed before he died.

Ye angels 'round the throne of grace,
Look down upon the braves,
Who fought and died on Shiloh's plain,
Now slumbering in their graves.
How many homes made desolate,
How many hearts have sighed.
How many like that drummer boy,
Who prayed before he died.

Camaraderie Between Both Sides

It wouldn't be too far of a stretch to think that having the chance to interact with someone your own age would be a welcome opportunity for these boys among men. After a battle, drummer boys from opposing sides might exchange glances or gestures of recognition, acknowledging the shared experience of surviving the combat. The shared language of drumming could create a subtle connection, even if they didn't directly interact, as the rhythms and melodies played by both sides might be familiar to each other. While there might have been instances of informal interactions, direct communication or fraternization was generally discouraged and considered inappropriate during wartime.

This letter, written by Union drummer Delavan Miller, shows the empathy that one side could have for the other.

After the fighting at Sailor's Creek had ended, Delavan Miller and his friends in the drum corps of the 4th New York Heavy Artillery found a Confederate drummer boy, wounded and taken prisoner. "My sympathies were stirred as they had never been before," Delavan recalled when he wrote his memoirs, "as a boy, scarcely 16 years old, was lifted out of the wagon.... He, too, was a drummer boy and had been wounded two or three days before. We got our surgeon and had his wound dressed and gave him stimulants and a little food, but he was... "all marched out," he said... We bathed his face and hands with cool water... [and] before leaving "Little Gray", as we called him, two boys knelt by his side and repeated the Lord's prayer... In the morning the little Confederate from the Palmetto state was dead and we buried him on the field with his comrades. Twas war- real genuine war."

Diary Recollections

Elisha Stockwell of Alma, Wisconsin:

We heard there was going to be a war meeting at our little log school house. I went to the meeting when they called for volunteers, Harrison Maxon (21), Edgar Houghton (16), and myself, put our names down.... My father was there and objected to my going, so they scratched my name out, which humiliated me somewhat. My sister gave me a severe calling down...for exposing my ignorance before the public, and called me a little snotty boy, which raised my anger. I told her, 'Never mind, I'll go and show you that am not the little boy you think I am.'

The Captain got me in by lying a little, as I told the recruiting officer I didn't know just how old I was but thought I was eighteen. He didn't measure my height, but called me five feet five inches high. I wasn't that tall two years later when I re-enlisted, but they let it go, so the records show that as my height.

I told her [his sister] I had to go down town. She said, "Hurry back, for dinner will soon be ready." But I didn't get back for two years.

Thomas Galwey of the Eighth Ohio Regiment:

There was considerable delay in issuing us clothing and equipment. It was not until the second week of [1861] that we were issued wooden guns, wooden swords and cornstalks with which to drill and mount guard. We went to parade in our shirts, still not being fully uniformed.

Elisha Stockwell after the Battle of Shiloh:

As we lay there and the shells were flying over us, my thoughts went back to my home, and I thought what a foolish boy I was to run away to get into such a mess I was in. I would have been glad to have seen my father coming after me.

Thomas Galwey on the horror of war:

The rains have uncovered many of the shallow graves. Bony knees, long toes, and grinning skulls are to be seen in all directions. In one place I saw a man's boot protruding from the grave...leaving the skeleton's toes pointing to a land where there is no war.

Charles W. Bardeen with the First Massachusetts Regiment:

There was no question of getting back to the reg¬iment .... I could see that my division was preparing to march, and while I did not actually run I certainly walked fast to get to it. It is curious how little annoy¬ances will keep themselves prominent even in time of danger. I had on thick woolen drawers which had somehow broken from the fastening that held them up. It was a warm day and as I hurried up the hill those drawers kept slipping down till they drove me almost distracted, disturbing my equanimity more than the danger did.

The Drums

Drums were made primarily in the important industrialized centers of the Northeast: Boston, New York and Philadelphia. There were no standards for drum construction but the vast majority of them measured 15"-16" in diameter and were 10"-12" deep. The shells were usually made of ash, maple or similar plyable woods. Wooden hoops were used to reinforce the drum which was "tuned" by adjusting ropes that crisscrossed around the shell and provided tension on calfskin or sheepskin heads. The four strand snare was constructed from a bronze hoop-mounted strainer with a leather anchor. Each drum featured a custom paintjob that made them ornamental.

According to DRUM magazine:

The crowning glory of many of these drums was their hand-painted decorations. Normally the drummer boy would receive his drum with the painting on the shell of the drum. Although there were no standards, a blue background was designated for an infantry unit, while a red background signified artillery. An American bald eagle most commonly emblazoned the Federal Army drums but sometimes the Confederates used it as well. Federal drums were also decorated with 13 stars for each of their 13 states. Confederate states were represented with 11 stars. With these beautiful decorations, it is no wonder that these drums were treasured long after the passionate sentiment of America's bloodiest battle had abated.

Although most drums from that era are preserved in museums, Civil War drums still exist on the market as antiques. One can expect to pay up to $7,500+ for one in good condition. A quick look on eBay reveals the high cost for original drums. Regardless, to own an original Civil War drum is to possess a piece of history.

Drumming Signals

Drummers on both sides of a conflict used their instruments to communicate in camp and on the field of battle. These signals were used for issuing duties and maneuvers.

Turn Right

Turn Left

Turn About

Long Roll

Water Call

Wood Call

First Sergeant's Call

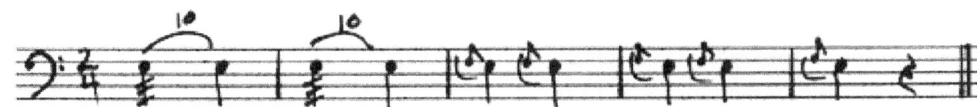

All Non-Commissioned Officers Call

Quotes on the Drummer Boy

"I can fight with ten times more spirit, hearing the band play some of our national airs, than I can without the music." – soldier to the Sanitary Commission

"It required but a glance at the countenances of the men to enable me to read the settled determination with which they undertook the task before them. The enemy, without waiting to receive the onset, broke in disorder and fled." – Gen. George Armstrong Custer on making his drummers play 'Yankee Doodle'

"For a marching column there is nothing like martial music of the good ol fashioned king." - Confederate drummer boy Delavan Miller

"I don't believe we can have an army without music." - General Robert E. Lee

"They please and interest the great majority of the soldiers and the men are almost universally proud of their band." - Union Secretary of the Sanitary Commission, Fredrick L. Olmstead

"The removal of wounded from the firing line was much more promptly and efficiently performed by the musicians than the 'ambulance corps.'" - General William Babcock Hazen

"The wounded men had little to interest them in their recuperation until the band of the 9th . . . marched from camp to play." – Union physician

"The boys think of the band as the important element of the army." - Charles W. Bardeen

"In North and South alike, small boys by the hundreds, and then by the thousands, eagerly enlisted and marched off to war like veterans." -   Webb Garrison in his book *A Treasury of Civil War Tales*

"At one point amidst the smoke and din, the men observed a white-haired drummer boy, dressed in Union blue, running toward their breastworks. Panic was etched on his face as he dove for the safety of Nutt and his men amid a shower of musketry. The boy made it to safety." - Captain E. E. Nutt

"The little fellow, becoming weak from the loss of blood looked up and said, "Caliber 68," and as he tottered he was seized by two soldiers and carried to the rear. I went up to my father…and found to my surprise that his eyes were suffused with tears of sympathy for the brave boy." – Remarks on the siege of Vicksburg, from twelve-year-old Fred Grant

"I never fired a shot. I was still a drummer boy. During much of that battle I served in the Medical Corps. Shot and shell and the screams of dying men and boys filled the humid air. A non-com told me to put away my drum." - James M. Lurvey recalling the Battle of Gettysburg

Veteran Drummers

Many of the drummer boys that survived the war participated in veteran reunions which were held every year. They would often march in parades and play the very same cadences that they played during the war.

In 1913 all honorably discharged veterans were invited to the reunion of the Battle of Gettysburg, drawing more than 50,000 members of the Grand Army of the Republic (the north) and the United Confederate Veterans (the south). Fifty years after the battle, many were in their 70s.

War veteran Flint Hartshorn joined Company C of the 12th Iowa Infantry in October 1861, and six months later found himself in the thick of the action at the Battle of Pittsburg Landing. According to one story of the fight, Hartshorn pulled his lieutenant from beneath a fallen horse and helped him to safety. About this time, Hartshorn suffered a wound that destroyed one eye, and impaired the other. He left the regiment with a disability discharge a few months later. Hartshorn married and settled in Dundee, Mich., where he and his drum became a regular at Grand Army of the Republic parades. He is pictured here towards the end of his life wearing G.A.R. medals. He died in 1919 at age 81.

Captain R.D. Parker, age 90, who played a drum at Lincoln's inauguration, as he took part in the final parade of the Grand Army of the Republic in Washington, D.C., closing the 70th annual encampment. The Grand Army of the Republic was an organization founded in 1866 for veterans of the Civil War.

Drummer Boy Gravesites

The grave marker of Henry Burke, a 13-year-old drummer boy who played a crucial part in the nearby Battle of Shiloh.

The grave marker of Paul Wilson, an escaped slave, was part of Company D of the United States 33rd Colored Troops Regiment.

The grave marker of drummer boy Edward Black (1853-1871) at Crown Hill Cemetery in Indianapolis, Indiana.

The grave marker of William H. Horsfall who saved the life of a wounded officer lying between the lines during the siege of Corinth.

The grave marker of Bernard Ross, who enlisted in the Union Army at the raw age of 12.

The grave marker of Johnny Hendricks who was a private in the 25th Iowa Volunteers.

Their Legacy

The legacy left behind by the Civil War Drummer Boy is one of courage. It is the story of boys who went off to war and returned home as young men. The suffering they endured was equal to the hardships experienced by the adults they served. Their sacrifice was that of a soldier. Their service, whether playing cadences on the march, reveille in camp, or orders on the battlefield was essential to the efficient functioning of an army on campaign. The drummer boy's skill and bravery cannot be overlooked. He volunteered to leave the safety of his home to serve in a war he did not begin, and in some cases, understand. When we remember the veterans who fought in the Civil War that divided our nation we cannot forget the contributions of the drummer boy.

Perhaps Ray Bradbury summed it up best in his story *The Drummer Boy of Shiloh* when his character the General confidently said to his drummer, *"Now, boy, you are the heart of the army. Think of that. You're the heart of the army."*

9th Vermont Infantry on the march

Sources

Print:

Albert A. Nofi, *A Civil War Treasury: Being a Miscellany of Arms and Artillery, Facts and Figures, Legends and Lore,* (Da Capo Press, 1992)

Bruce & Emmett Drummers' and Fifer's Guide (1862)

Carolyn Reeder, "Drummer boys played important roles in the Civil War, and some became soldiers" (*Washington Post*, February, 2012)

Charles Stewart Ashworth, *A New Useful and Complete System of Drum Beating (January, 1812)*

Chet Falzerano, "Historic Collectible: Civil War Drums" (*DRUM! Magazine*, December, 2011)

Elizabeth M. Collins, "The Beats of Battle: Images of Army Drummer Boys Endure" (*Soldiers: The Official U.S. Army Magazine*, December, 2013)

Gardiner A. Strube, *The Rudimental Principles of Drum – Beating (1870)*

Marcie Schwartz, "Children on the Battlefield: Heroism and Sacrifice" (civilwartrust.org)

Max Louis Rossvally, *Charlie Coulson a Drummer Boy: A True Story in the American Civil War* (188?)

Meserette Kentake, *"Alexander H. Johnson: The first drummer boy" (Kentake Page, July 2015)*

Robert H. Hendershot, "Drummer Boy of the Rappahannock" (*National Tribune*, Grand Army of the Republic (GAR), July 1891)

Robert J. McNamara, "Civil War Drummer Boys" (history1800s.about.com)

Ronald S. Coddington, "Colonel Shaw's Drummer Boy" (*The New York Times*, March, 2013)

The Indiana Democrat, Thomas Hubler (Indiana, Pennsylvania, November, 1883)

U.S. Civil War History & Genealogy, "The Drummer Boys" (genealogyforum.com)

Websites:

American Battlefield Trust

AmericanCivilWar.com

American Civil War Magazine

Civil War Monitor

DRUM Magazine

Emerging Civil War

Fredericksburg Spotsylvania National Military Park

History.net

Library of Congress Archives

Modern Drummer Magazine

Off Beat Blog

Sons of the South

Time Life Civil War Series

About the Author

Michael Aubrecht is an author, drummer and historian from Fredericksburg, Virginia. He is the author of several books to include *The Civil War in Spotsylvania County, Historical Churches of Fredericksburg, Thomas Jefferson and the Virginia Statute for Religious Freedom* and the best-selling *FUNdamentals of Drumming for Kids*. Michael has been a contributing writer for *Drumhead, Modern Drummer, Patriots of the American Revolution* and *America's Civil War* magazines. He maintains a popular blog titled *The Naked Historian* at www.michaelaubrecht.wordpress.com. Contact Michael at ma@pinstripepress.net.

About the Foreword Author

Daniel Glass is an award-winning drummer, author, historian and educator. He is widely recognized as an authority on classic American drumming and the evolution of American Popular Music. A member of the pioneering swing group Royal Crown Revue, Daniel has recorded and performed with many top artists. As an educator, Daniel has published five books and three DVDs, including the award-winning titles *The Century Project, The Roots of Rock Drumming,* and *The Commandments of Early Rhythm and Blues Drumming*. He is a regular contributor to publications like *Modern Drummer, DRUM* and *Classic Drummer*. Visit Daniel online at www.danielglass.com.

Acknowledgements for this Project

I would like to thank Daniel Glass for his outstanding Foreword. I can't think of anyone more qualified to comment on the history of drumming in any capacity. I also want to thank my friends Rich Redmond and Scott Mingus for their kind quotes endorsing this project. I want to thank all of the historians and drummers who continue to inspire me to do better work both on the page and behind the drum set. I also want to thank all of the drummer boys of the past who left behind their pictures and their words so that we may better understand the experiences that they endured during America's Civil War.

Also by Michael Aubrecht

From 1861 to 1865, hundreds of thousands of troops from both sides of the Civil War marched through, battled and camped in the woods and fields of Spotsylvania County, earning it the nickname 'Crossroads of the Civil War.' Focusing specifically on the local Confederate encampments, renowned author and historian Michael Aubrecht draws from published memoirs, diaries, letters and testimonials from those who were there to give a fascinating new look into the day-to-day experiences of camp life in the army.

Historic Churches of Fredericksburg: Houses of the Holy recalls stories of rebellion, racism and reconstruction as experienced by Secessionists, Unionists and the African American population in Fredericksburg's landmark churches during the Civil War and Reconstruction eras. Using a wide variety of materials compiled from the local National Park archives, author Michael Aubrecht presents multiple perspectives from local believers and nonbelievers alike who witnessed the country's Great Divide.

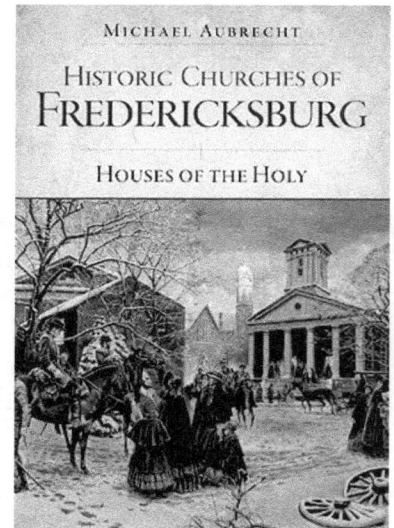

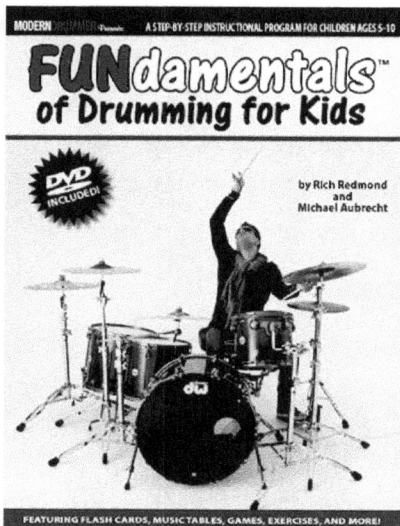

Co-written by drummers/authors Rich Redmond and Michael Aubrecht, FUNdamentals of Drumming for Kids is the culmination of the collective experiences and educations of a professional player and a player/parent who both understand the tangible benefits of exposing children to music at a young age. Whether or not a child decides to pursue an instrument seriously, the skill set he or she develops will provide an edge in all aspects of education.

About the Book:

As a passionate drummer himself, author Michael Aubrecht's "The Long Roll" is the ultimate look at the historical and cultural importance of wartime drummers. This book is surely the most informative and entertaining look at the subject.

- Rich Redmond, Drummer for Jason Aldean and author of *C.R.A.S.H. Course for Success: 5 Ways to Supercharge Your Personal and Professional Life*

Many drummers, including my own great-great-grandfather, were mere teenagers during the war. Michael Aubrecht neatly provides an overview of the role of drummers in the war and relays several interesting anecdotes about famous and less well-known drummers in this fascinating new account.

- Scott Mingus, author of over a dozen Civil War books including *Gettysburg Glimpses: True Stories from the Battlefield* and *Human Interest Stories of the Gettysburg Campaign*

About the Title:

"The Long Roll" called troops to arrange
themselves in line prior to an attack.

www.ingramcontent.com/pod-product-compliance
Lightning Source LLC
Chambersburg PA
CBHW081341090426
42737CB00017B/3235